Disasters

Level 7 – Turquoise

Helpful Hints for Reading at Home

The graphemes (written letters) and phonemes (units of sound) used throughout this series are aligned with Letters and Sounds. This offers a consistent approach to learning, whether reading at home or in the classroom.

HERE IS A LIST OF PHONEMES FOR THIS PHASE OF LEARNING. AN EXAMPLE OF THE PRONUNCIATION CAN BE FOUND IN BRACKETS.

Phase 5			
ay (day)	ou (out)	ie (tie)	ea (eat)
oy (boy)	ir (girl)	ue (blue)	aw (saw)
wh (when)	ph (photo)	ew (new)	oe (toe)
au (Paul)	a_e (make)	e_e (these)	i_e (like)
o_e (home)	u_e (rule, cube)		

Phase 5 Alternative Pronunciations of Graphemes			
a (hat, what)	e (bed, she)	i (fin, find)	o (hot, so, other)
u (but, unit)	c (cat, cent)	g (got, giant)	ow (cow, blow)
ie (tied, field)	ea (eat, bread)	er (farmer, herb)	ch (chin, school, chef)
y (yes, by, very)	ou (out, shoulder, could, you)		

HERE ARE SOME WORDS WHICH YOUR CHILD MAY FIND TRICKY.

Phase 5 Tricky Words			
oh	their	people	Mr
Mrs	looked	called	asked
could			

TOP TIPS FOR HELPING YOUR CHILD TO READ:

- Allow children time to break down unfamiliar words into units of sound and then encourage children to string these sounds together to create the word.

- Encourage your child to point out any focus phonics when they are used.

- Read through the book more than once to grow confidence.

- Ask simple questions about the text to assess understanding.

- Encourage children to use illustrations as prompts.

PHASE 5 /a/

This book focuses on /a/ and the alternative pronunciations of its grapheme. It is a Turquoise level 7 book band.

Can you sort these words into two groups?
One group has a as in **mat**.
One group has a as in **bagel**.

apricot

lady

cat

catch

bacon

last

angel

plaster

We live on an amazing planet. From parks to plains to wetlands, there are all sorts of sights to see. Fantastic people, plants and animals are found across the globe.

However, there can sometimes be frightening events that happen on this planet. What happens when the rain will not stop? What happens when the land beneath us falls apart?

On some days, you might feel a lot of wind outside. However, a hurricane is quite a lot more frightening. A hurricane is a swirl of wind that can destroy whatever stands in its path.

Hurricanes start out at sea. The winds swirl around and gain speed as they get closer to land. As they hit the land, they may pass over towns.

Hurricane winds are so fast that they will not just rip a branch off a tree, they will rip up the entire tree!

You must stay safe if a hurricane is on its way. Do not wander in the street or stand and watch. You must get to a safe shelter.

Disaster can strike when you least expect it. Sometimes, the ground is not as safe as we think it is.

A landslide is when part of a hill or cliff drops away. When this happens near a town, it can squash things underneath it, such as people's homes.

The part of our planet under the ground is made of rock. When that rock shifts under the sea, it can make a colossal wave.

If the wave lasts until it meets the land, it can destroy all that lies in its path. The waves can be so vast that entire towns can be washed away.

Each time that a disaster happens, lots of people come to help the people affected. Experts might go to help people repair their homes.

It is important to look after people, as disasters can affect us all. We need to make sure that help is there when they happen.

©2023 BookLife Publishing Ltd.
King's Lynn, Norfolk, PE30 4LS, UK

ISBN 978-1-80505-107-7

All rights reserved. Printed in China.
A catalogue record for this book is available from the British Library.

Disasters
Written by Louise Nelson
Designed by Lucy Otter

FSC
www.fsc.org
MIX
Paper from responsible sources
FSC® C113515

An Introduction to BookLife Readers...

Our Readers have been specifically created in line with the London Institute of Education's approach to book banding and are phonetically decodable and ordered to support each phase of the Letters and Sounds document.

Each book has been created to provide the best possible reading and learning experience. Our aim is to share our love of books with children, providing both emerging readers and prolific page-turners with beautiful books that are guaranteed to provoke interest and learning, regardless of ability.

BOOK BAND GRADED using the Institute of Education's approach to levelling.

PHONETICALLY DECODABLE supporting each phase of Letters and Sounds.

EXERCISES AND QUESTIONS to offer reinforcement and to ascertain comprehension.

CLEAR DESIGN to inspire and provoke engagement, providing the reader with clear visual representations of each non-fiction topic.

AUTHOR INSIGHT:
LOUISE NELSON

Over her time as a children's author, Louise Nelson's natural curiosity about the world has inspired her to be a lifelong learner. She is eager to pass on whatever she learns with the next generation of inquisitive young minds. Louise's wide range of interests has helped her hone a talent for producing extremely versatile work and motivates her to always be on the lookout for fascinating new topics to explore.

PHASE 5 /a/

This book focuses on /a/ and the alternative pronunciations of its grapheme. It is a Turquoise level 7 book band.

Image Credits Images are courtesy of Shutterstock.com. With thanks to Getty Images, Thinkstock Photo and iStockphoto. Cover – 1Arts, Belozersky, IgorZh, KurArt, Thara1979. 4–5 – arzmadani, John D Sirlin. 6–7 – Felix Mizioznikov, Vladislav Gurfinkel. 8–9 – Jewelzz, MyImages – Micha. 10–11 – mrizag, NayaDadara. 12–13 – Christian Vinces, Youkonton. 14–15 – 4.murat, Microgen.